Electrons in Matter and Link to Redox Chemistry

By Malika Ammam, PhD

Copyright© 2017 Malika Ammam. All rights reserved.

Discount Offers

5% OFF of the book price for purchases of 1-5 books.

8% OFF of the book price for purchases of more than 5 books.

To receive the discount money, send your request through https://www.malika-ammam.com/ with your order details and PayPal account. Make sure that your order details (amazon or other sites) passed the 30 days return policy.

Thank you,

Introduction

As a teacher of physical chemistry, I noticed that students, even in advanced classes, have difficulties in understanding the basics of redox chemistry. In this Section 2, I attempted to summarize some fundamental principles related to electrons, including the atomic structure, periodic table, main electronic models, electronic configuration, and the energy levels of electrons. The identification of the electronic configurations of elements and visualization of electrons in their outer shells are very important concepts in understanding redox processes. To further clarify the discussed concepts, numerous questions and problems with detailed answers are provided. Most of these questions are formulated by students like you. I believe that this Section 2 would greatly help students with levels varying from high school to advanced university classes.

Abstract

Keep in mind that redox reactions are chemical processes but not all chemical reactions are redox. Redox chemistry, also known as electrochemistry, involves the loss or gain of electrons and could track the movement of consumed or released electrons during the process. Electrons are believed to exist since the early stages of formation of the Universe during the Big Bang. These subatomic particles, called electrons, are part of atoms, molecules, and matter in general. To better understand electrochemical processes discussed in later sections, some basics related to electrons are given here, including the structure of atoms, periodic table, main electronic models, electronic configuration, and the energy levels of electrons.

1. **Atoms as building blocks of matter**

1.1. **Matter and substances**

Matter concerns everything that physically exists with mass and could occupy a space. It is formed of atoms, assembled in a way to give it proper physical and chemical properties[1-2]. Atoms could not be broken down or destroyed by ordinary chemical procedures. All discovered atoms are gathered in a table, called the periodic table of elements, and examples include hydrogen (H), oxygen (O), zinc (Zn), gold (Au), uranium (Ur), among others. Matter could be assembled by the same exact atoms (e.g., pure gold, iron, copper) or by a mixture of different atoms (e.g., water, plastic, glass).

At the macroscopic level, substances composed of one type of matter with distinct composition and distinct properties are called pure substances. Examples include various pure metals (e.g., gold, copper, silver) or pure gases (e.g., hydrogen, oxygen, nitrogen). These pure substances cannot be separated by physical processes, such as filtering, centrifugation, and distillation. Pure substances may also contain compounds composed of two or more elements (e.g., pure water, pure olive oil, and pure ammonia). Unlike the first category, these substances can be broken down into their constitutional elements or atoms using chemical processes, such as elemental analysis.

Two or more substances of different proportions form mixtures. Unlike pure substances, mixtures could be separated by means of physical processes, such as filtration, distillation and/or centrifugation to yield separate compounds. Depending on the particle size of each compound, mixtures could be categorized as homogenous or heterogeneous. Homogenous mixtures have uniform compositions and properties in terms of color and distribution of solute/solvent throughout the mixture (e.g., HCl mixed with water). By contrast, heterogeneous mixtures have different distributions of the constituents in the mixture (e.g., olive oil mixed with water). This facilitates the separation of heterogeneous mixtures using physical processes to identify the main constituents (e.g., separating olive oil from water).

1.2. Elements and structure of atoms

An element is made of a network of the same atoms[3-7]. For example, 99% gold rod is made of 99% gold atoms (Au) and the other 1% are impurities, which could be copper (Cu), iron (Fe), among others. Atoms of various elements have different structures, thus different weights and chemical properties. Note that everything on earth so far analyzed, is made of one or several atoms. Unlike substances or compounds which could be destroyed and transformed into other forms, atoms are indestructible but their structures could be altered.

From the structural viewpoint, atoms are made of the same components: a positively charged nucleus in the center surrounded by negatively charged electron(s) to form a neutral unit[3-5]. The nucleus contains two subatomic particles: protons and neutrons with almost similar masses. The neutrons are electrically neutral with no change (0) and the protons are positively charged (+1). The masses of the neutron and proton are respectively 1.00728 amu and 1.00867, where amu represents the standard mass unit of $1.660539040(20) \times 10^{-27}$ Kg, meaning that there are $6.022140857 \times 10^{23}$ (or Avogadro's number) in 1 amu.

Electrons are negatively charged subatomic particles turning around the nucleus at high velocities. The mass of an electron is 0.000549, which is 1836 lighter than that of the nucleus and its charge is -1. The negatively charged electrons are bound to the positively charged protons of the nucleus by electromagnetic forces. In stable ground states, the total number of protons is the same as the electrons and overall charge of the atom is neutral. However, electrons could be ejected from or added to atoms. Keep in mind that the electrons involved in chemical and redox reactions are those occupying the top outer shells (or valence shells), which mainly determine the chemical properties of atoms and substances.

The number of protons in the nucleus is referred to as the Z number or atomic number, which is the same as the number of electrons in neutral atoms. Note that Z is very important in redox reactions and often used to express the atomic charge in terms of loss or gain of electrons. The mass number (A) of an atom is defined as: $A = Z + N$, where Z is the atomic number (or number of protons) and N is the number of neutrons.

1.3. *Some characteristics of atoms*

Atoms are characterized by an atomic radius (radii), measured from the center of the nucleus to the boundary of the top electronic shell[3-9]. For simplification reasons, atoms are often considered as spheres but the structure is more complex. Because electrons are constantly moving around the nucleus at high velocities, they have no definite positions and consequently their presence is expressed in terms of probability of distribution in atomic orbitals or subshells. The atomic size varies between 0.3-3 angstrom (Å), which is about 10^5 times higher than the nucleus radii.

Atoms are characterized by electronegativities, defined as the ability of the atom to attract electrons towards itself[3-7,10-11]. Electronegativity depends on both the atomic number Z and the distance separating the nucleus from the valence electrons. High numbers of protons and shorter distances to the valence shell should induce elevated electronegativities. Atoms with high electronegativity values easily attract electrons to their valence shells. By contrast, atoms with low electronegativities have more tendencies to lose electrons from their valence shells when they interact with other atoms with higher electronegativities. Note that electronegativity is a property more associated with an atom in a molecule rather than in atom alone, but electron affinity and ionization energy are more appropriate for separate atoms.

Atoms could also be distinguished by their ionization energies (or ionization potentials), defined as the amount of energy required to remove the most weakly bound electrons from the valence shell to form cations[3-7,12]. Each electron in an atom is characterized by its own ionization energy. Electrons close to the nucleus have higher ionization energies because they are strongly bound by electrostatic forces to the protons of the nucleus. By contrast, electrons located in the valence shell with appreciable distances from the nucleus are easy to eject (remove or oxidize), hence their ionization energies are low. The ionization potential is very useful in electrochemistry since it describes the energy required for removing one electron during oxidation.

2. Periodic table of elements

The first classification of elements was proposed in 1869 by ''Dmitri Mendeleev'', and was based on increasing atomic masses to form columns of elements with similar physical and chemical properties [4-6,13]. The modern periodic table uses the same principle but more advanced and contains more discovered elements. As of 2016, a total of 118 different elements have been discovered, ranging from the simplest ''hydrogen'' to most complex know elements to date "oganesson". Each element is represented by a unique abbreviation symbol, which does not necessarily match the name in English due to their various origin of discovery. For example, gold discovered during 3000 BC, was named in Latin "Aurum", and hence was symbolized as "Au". Similarly, Iron comes from Latin "Ferrum" and given the symbol, "Fe". These elements are distributed in groups ranging from 1 to 18 and rows (or periods) from 1 to 7, with two additional rows representing the *lanthanide* and *actinides*.

The elements of the periodic table could be organized according to their subshell electronic configurations, which determine their chemical properties. In the periodic table, the elements could be identified through groups, periods or even block. Elements of the same column (group) share the characteristics of having the same electronic configuration of the valence shell. They are organized according to increased atomic number from top to bottom of the periodic table. However, some parts of the periodic table do not necessarily follow this trend, such as the *d*-block and *f*-block, where similarities appear when moving along the horizontal than the vertical direction. The most familiar groups are alkali metals (group 1), alkaline earth metals (group 2), halogens (group 17), Noble gases (group 18), and transition metals (group 3 to 12).

Elements of the same group show increasing atomic radius from top to bottom. Some exceptions to this rule exist when atomic contraction occurs due to strong attractive forces between the electrons and nucleus, such as in some transition metals. The ionization energy reduces from top to bottom in a period because of the increase in atomic radius. Also, the more the valence electrons are far from the nucleus, the more they become less bonded, making them easier to eject (or oxidize) using lower ionization energies. The increase in distance between the valence electrons and nucleus, from top to bottom in a period, also reduces the electronegativity since the electrons become farther from the nucleus. Some exception might also apply to this rule.

Periods can also be used to identify and compare elements in the periodic table, especially for groups failing to show reasonable trends. Periods are particularly more useful to compare the lanthanides and actinides elements as the horizontal trend makes more sense than the vertical (group). Similar to groups, the chemical properties can also be compared following periods, including atomic radius, ionization energy, and electronegativity. From left to right, the atomic radius decreases, thus ionization energy and electronegativity increase.

The periodic table might also be divided into blocks, each gathering a certain number of elements organized according to their valence electrons. Each block contains elements with electrons occupying the same subshell or outer orbital, namely *s*, *p*, *d*, and *f* : *s*-block (first two groups), *p*-block (last six groups), *d*-block (groups 3 to 12 or so-called transition metal block), and *f*-block (lanthanides and actinides).

3. **Basic electronic models of atoms**

As mentioned above, atoms are made of positively charged protons and neutral neutrons occupying the nucleus surrounded by negatively charged electrons moving around it at high speed. Atoms in the ground states have no charge since the number of electrons equals that of the protons. However, atoms can be the subject of chemical reactions to lose one or several electrons to form positively charged species (cations) or gain one or several electrons to form negatively charged species (anions). Both processes of losing and gaining electrons are basic to redox chemistry (or electrochemistry).

Because electrons have very small masses than the nucleus (1/1800) and absolute charge equals to that of protons, both parts are attached to each other by coulombic attractive forces. Electrons occupy no definite positions around the nucleus but they move around it at high speeds, forming electronic clouds. Their positions at time *t* are thus expressed in terms of probability of presence in an energy level or orbital.

Basically, there are seven energy levels matching the seven periods (rows) of the periodic table, which could be occupied by electrons in their ground states (Bohr model)[4-7,14-15]. These energy levels are quantified, meaning that electrons cannot occupy spaces between these definite levels. The valence (or outermost) electrons fill in the energy level corresponding to the period number. Each of the seven energy levels could accept a maximum number of electrons summarized in Table 1.

Table 1: Maximum number of electrons which could occupy each energy level.

Energy level	1	2	3	4	5	6	7
Maximum number of electrons	2	8	18	32	50	72	98

Electrons in each energy level are distributed in different sublevels called orbitals (*s*, *p*, *d*, *f*, *g*), from low to high energy. The *s* orbital could hold up to two electrons. The *p* orbital is divided into three different *p* sub-orbitals, each with a maximum of two electrons. The *d* and *f* orbitals are divided respectively into five and seven different sub-orbitals, each could hold up to two electrons. The electrons are distributed in these orbitals in a predetermined pattern starting from the lowest energy orbital *s* to the highest possible.

4. Electronic structure

4.1. Quantum numbers

The position of each element in the periodic table often follows its electronic configuration. The quantum theory based on solutions of the Schrödinger's Equation could be used to fill in the electronic shells and determine the electronic configuration of elements [14-17]. The Schrödinger model proposes four quantum numbers, namely *n*, *l*, *m*, and *s*. The first three numbers describe the orbital occupied by the electron and the fourth represents the characteristics of the electron in the orbital in terms of orientation (or spin).

n is the principal quantum number with positive integers (e.g., 1, 2, 3, etc.), measuring the distance of an orbital from the nucleus. Orbitals having the same *n* are contained in the same shell, and the highest *n* for any element corresponds to the number of its row in the periodic table. The maximum number of electrons that could occupy the nth energy level is $2n^2$.

l is the azimuthal quantum number characterizing the shape of the orbital and its angular momentum. *l* could have integers from 0 to (n - 1). Orbitals having the same *n* and *l* values are contained in the same subshell, each could be filled with up to $(4l + 2)$ electrons. The *l* values of 0, 1, 2, 3 and 4 correspond to subshells designated by *s*, *p*, *d*, *f* and *g*, respectively.

m is the magnetic quantum number with integer values ranging from $-l$ to *l*, representing specific orbitals within subshells. A shell containing one orbital *s* will have an *m* value of 0. Shells containing *p* orbitals will always divide into 3 different *p* sub-orbitals, with corresponding *m* values of -1, 0, and 1. Each sub-orbital could hold up to a maximum of two electrons.

s is the spin quantum number representing the intrinsic angular momentum. The maximum two electrons occupying each orbital might be viewed as tiny magnets spinning around their axis in clockwise or counterclockwise directions. Thus, each electron in the orbital could either have the value of $-1/2$ or $+1/2$, corresponding respectively to down and up orientations.

In sum, the *s* subshells could hold up to ($4 \times 0 + 2 = 2$ electrons), *p* subshells up to ($4 \times 1 + 2 = 6$ electrons), and *d* sub-shell up to ($4 \times 2 + 2 = 10$). For all subshells, the *s* could have the values of −1/2 or +1/2. In an atom, the Pauli Exclusion Principle indicates that each electron must have a different set of four quantum numbers. Table 2 lists a summary of the four quantum numbers for n = 1, 2, and 3.

Table 2: Summary of the quantum numbers for n = 1, 2, and 3.

n	1	2		3		
l	0	0	1	0	1	2
Subshell	1s	2s	2p	3s	3p	3d
m	0	0	−1, 0, 1	0	−1, 0, 1	−2, −1, 0, 1, 2
Number of orbitals in subshell	1	1	3	1	3	5
Maximum electrons in subshell	2	2	6	2	6	10
Electron spin	±1/2	±1/2	±1/2	±1/2	±1/2	±1/2

4.2. Energy levels and determination of electronic configuration

The energy levels of atoms containing several electrons are determined by both quantum numbers, n and l[14-17]. The simplified diagonal arrow method shown in scheme 1 could be used to determine the order of the energy levels.

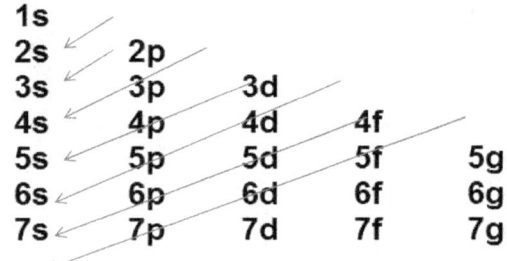

Scheme 1: Simplified representation of the diagonal arrow method.

The above energy level diagram is the simplest way to determine the electronic configuration of atoms following the order: $1s < 2s < 2p < 3s < 3p < 4s < 3d < 4p < 5s < 4d < 5p < 6s < 4f < 5d < 6p <$, etc. The total number of electron in each atom should be distributed on the different orbitals from the lowest to the highest in energy, by specifying each orbital and maximum number of electrons. For example, the electronic configuration of oxygen (total of 8 electrons) should be written as: $1s^2 2s^2 2p^4$. This means that the first orbital contains 2 electrons, the second 2, and the third 4.

For atoms with large numbers of electrons, the electronic configurations become long. These could be simplified by emphasizing on the outermost electrons involved in chemical reactions. The procedure consists of writing first the previous element with full subshell, often a noble gas between brackets, then adding the remaining outer electrons. For example, the full electronic configuration of Cr is: $1s^2 2s^2 2p^6 3s^2 3p^6$. This can be simplified by using the noble gas $_{18}Ar$, followed by the remaining outer shell electrons to yield: $[Ar]4s^2 3d^4$. Note that the electronic configurations are very useful in electrochemistry to figure out losses or gains in electrons during redox reactions.

To better visualize the electronic configurations of elements, orbital diagrams could also be used to distribute the electrons on the different orbitals, often presented as boxes or circles. The filling of each orbital (presented here as boxes) with electrons must start with one electron in each box of the orbital with parallel spins (Hund's rule). Afterward, each box should be completed by a second electron with the opposite spin until exhaustion of the electrons. For example, in the configuration of Cr presented below, each box of the d orbital contains electrons with spins up, in accordance with the Hund's rule.

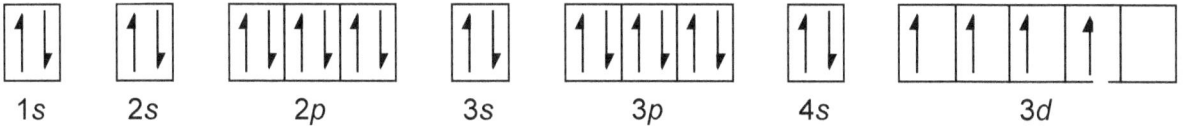

The energy of the orbitals increases from $1s$ to $3d$, conforming with the diagonal arrow method or as presented in Scheme 2 for Cr. Electrons in lower orbitals are more stable than those occupying the outer shell which could participate in chemical reactions, such as redox processes.

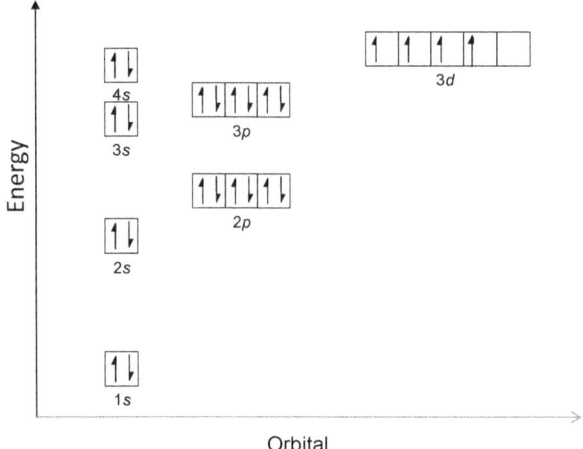

Scheme 2: Energy diagram of Cr from the lowest to the highest orbitals.

Summary

Redox (or electrochemical) reactions involve the loss or gain and transfer of electrons. These electrons are part of atom structure. Electrons are negatively charged and bond to the positively charged nucleus by electrostatic forces. Because the mass of electrons is very small than that of the nucleus, they turn around it at high speeds forming electronic clouds. The number of electrons, specifically those occupying the outer shell (or valence electrons), often determines the properties of atoms. Therefore, elements are classified in the periodic table according to their total number of electrons and resulting properties, such as electronegativity, ionization energy, and ionic radii. The elements are distributed in groups ranging from 1 to 18 and rows (or periods) from 1 to 7, with two additional rows for the lanthanide and actinides. Electrons in each atom are distributed on different energy sublevels designated by the orbitals (s, p, d, f, g), corresponding respectively to the azimuthal quantum number l (0, 1, 2, 3, 4). Each orbital could hold up to ($4 \times l + 2$) electrons and each electron could have a spin of $-1/2$ or $+1/2$. In an atom, the Pauli Exclusion Principle indicates that each electron must have different set of four quantum numbers. In sum, the total number of electrons of each atom could be distributed on the different orbitals from lowest to highest in energy (s, p, d, f, g) by specifying each orbital and maximum number of electrons that could hold. The simplest way to achieve this is by using the diagonal arrow method or energy level diagram, allowing quick identification of the electronic configuration of each element. Note that the electronic configurations and orbitals are very important concepts in redox processes.

References

1. 't Hooft, G. (1997). In search of the Ultimate Building Blocks. Cambridge University Press, page 6.
2. Matter (physics), McGraw-Hill's Access Science: Encyclopedia of Science and Technology Online.
3. IUPAC (ed.), Chemical Element, Gold Book.
4. Myers, R. (2003). The Basics of Chemistry. Greenwood Press. p. 85.
5. Earnshaw, A.; Greenwood, N. (1997). Chemistry of the Elements (2nd ed.).

6. Zumdahl, S. S. (2002). Introductory Chemistry: A Foundation (5th ed.). Houghton Mifflin.

7. Smirnov, B. M. (2003). Physics of Atoms and Ions. Springer. pp. 249-272.

8. Ghosh, D. C.; Biswas, R. (2002), Theoretical Calculation of Absolute Radii of Atoms and Ions. Part 1. The Atomic Radii, International Journal of Molecular Science, 3:87-113.

9. Dong, J. (1998). "Diameter of an Atom". The Physics Factbook.

10. IUPAC, Compendium of Chemical Terminology (1997), 2nd ed. (the "Gold Book").

11. Jensen, W. B. (1996), Electronegativity from Avogadro to Pauling: Part 1: Origins of the Electronegativity Concept, Journal of Chemical Education, 73 (1): 11-20.

12. IUPAC, Ionization potential, Gold Book.

13. Scerri, E. R. (2007), The Periodic Table: its Story and its Significance. Oxford University Press US. pp. 205-226.

14. Liboff, R. L. (2002). Introductory Quantum Mechanics. Addison-Wesley.

15. Griffiths, D. J. (2004). Introduction to Quantum Mechanics (2nd ed.). Prentice Hall.

16. Halzen, F.; Martin, A. D. (1984). Quarks and Leptons: An Introductory Course in Modern Particle Physics, John Wiley & Sons.

17. Peleg, Y.; Pnini, R.; Zaarur, E.; Hecht, E. (2010), Quantum Mechanics (2nd ed.), Schuam's Outlines, McGraw Hill (USA).

Section 2

Practical Questions and Problems with Solutions

A set of practical questions and problems with detailed solutions are provided to better understand the discussed concepts. The questions and problems range from simple to complex.

Q1. i) As of 2016, how many elements are so far discovered? ii) Each element is composed of which basic structural unit? iii) What is the smallest and simplest element?

Ans1. i) As of 2016, a total of 118 elements have so far been discovered. These elements are gathered in what is called the periodic table of elements. ii) All the elements are made by a basic structural unit called the atom. iii) The smallest and simplest element is called hydrogen and its symbol is H.

Q2. Let's consider computer laptops as a study subject. i) In your view, a laptop is composed of which elements? ii) What elements from the periodic table do you think exist in the laptop?

Ans2. The visible parts of a laptop are made of plastic and perhaps glass. The electronic components and circuits inside the laptop contain metals, such as gold and copper. Thus, a laptop is made of various elements.

ii) For example, plastic is a polymer backbone of carbon (C), hydrogen (H), and other atoms. Glass is made of silicon (Si) and oxygen (O). The microelectronics are generally made of gold (Au) and other elements like copper (Cu). These are just some examples but a laptop contains more elements in its composition.

Q3. i) Could the elements of the periodic table be displaced from their actual positions? ii) If so, why?

Ans3. i) The position of each element in the period is stable and unlikely to be displaced. ii) The elements are classified according to their structural features from the smallest atom to the biggest, as well as their chemical properties in terms of metallic, nonmetallic or semi-metallic.

Q4. i) In the periodic table, does the symbol of the element always match its current name? ii) Provide some examples.

Ans4. i) The symbol of each element does not always match the name of the element because these elements have different discovery origins in terms of cultural and language backgrounds. ii) For example, the symbol Fe is given for Iron and comes from Latin ''Ferrum''. However, other elements have symbols which match the name. These include Hydrogen (H), Tellurium (TE), and Iodine (I).

Q5. During your daily or weekly trip to the grocery store, what kind of atoms do you think are present at the store?

Ans5. The store contains a large number of items, including vegetable/fruits, dairy products, canned foods, cereals, storage containers, fridges/freezers, among other. Each of these items is made of atoms, molecules, and atomic/molecular networks. For example, vegetable/fruits contain vitamins, minerals, carbohydrates, among others. Minerals are made of Ca, Mg and Na, and vitamins of C, O, H, N, among others. Cereals, pasta and rice are made of carbohydrates (C, O, H, etc.). The glass contains networks of Si and O. The fridge is made of steeliness steel (Fe and Ni) and plastic of C backbone based polymers, among others.

Q6. i) What are the basic subatomic particles present in any given element of the periodic table? ii) What are the subatomic particles present in a hydrogen atom? iii) How far or close these subatomic particles are from each other in an atom? iv) Calculate the mass number A of the hydrogen atom. v) What will happen to the hydrogen atom if it loses an electron?

Ans6. i) Any atom is made of a nucleus containing protons and neutrons surrounded by electrons. ii) A hydrogen atom contains one proton in its nucleus (without neutrons) surrounded by one electron. iii) The neutrons and protons are close to each other but the electrons are distant from the nucleus. iv) The mass number of the hydrogen atom is: $A = N + Z = 0 + 1 = 1$. v) When a hydrogen atom loses the electron, the remaining atom will only contain a proton in its nucleus ($H - e^- \rightarrow H^+$). That is why hydrogen without an electron (H^+) is often called a proton.

Q7. i) Define a mole of any substance? ii) How many particles exist in a mole?

Ans 7. i) A mole of any substance contains the Avogadro's number. ii) There are $6.022140857 \times 10^{23}$ particles in one mole of any substance.

Q8. Consider nitrogen element in the periodic table. i) What is its abbreviation symbol? ii) How many protons, neutrons, and electrons are in the nitrogen atom? iii) What are the atomic number and mass number of nitrogen?

Ans8. i) The symbol of nitrogen is N. ii) It contains 7 protons and 7 neutrons in its nucleus. iii) Hence, $Z = 7$ and $A = Z + N = 7 + 7 = 14$.

Q9. i) Why are the electronic configurations important in electrochemistry? ii) Identify the ground-state electronic configurations of the following species: W, H^+, H^-, Cl^-, As, Co^{2+}, Cu, S^{2-}, Kr, and C. iii) For each species, identify which electrons could be lost through oxidation processes and why? The number of electrons of each element can be found in the periodic table.

Ans9. i) The knowledge of the electronic configuration allows determining which electrons could be lost or gained by a species. ii) and iii) Determining the electronic configuration of each species requires the knowledge of the total number of electrons.

W has 74 electrons, which have to be distributed on the different orbitals from lowest to the highest in energy. The diagonal arrow method helps to distribute these electrons: $1s < 2s < 2p < 3s < 3p < 4s < 3d < 4p < 5s < 4d < 5p < 6s < 4f < 5d < 6p...$

W: $1s^2 2s^2 2p^6 3s^2 3p^6 4s^2 3d^{10} 4p^6 5s^2 4d^{10} 5p^6 6s^2 4f^{14} 5d^4$

For cations, the total number of electrons is determined by subtracting the charge from the total number of electrons of the element. For example, the number of electrons in Co is 27, and Co^{2+} has 2 electrons less, thus 25.

For anions, the total number of electrons is determined by adding the charge of the ion to the total number of electrons of the element. For example, S has 16 electrons and S^{2-} has 2 additional electrons, thus 18 electrons in total.

Therefore, the electronic configurations of all the species are:

H^+ : $1s^0$

Cl^- : $1s^2 2s^2 2p^6 3s^2 3p^6$

As: $1s^2 2s^2 2p^6 3s^2 3p^6 4s^2 3d^{10} 4p^3$

Co^{2+} : $1s^2 2s^2 2p^6 3s^2 3p^6 4s^2 3d^4$

Cu : $1s^2 2s^2 2p^6 3s^2 3p^6 4s^1 3d^{10}$

S^{2-} : $1s^2 2s^2 2p^6 3s^2 3p^6$

Kr : $1s^2 2s^2 2p^6 3s^2 3p^6 4s^2 3d^{10} 4p^6$

C : $1s^2 2s^2 2p^2$

Q10. Consider the two oxygen isotopes ($^{18}_{8}O$ and $^{16}_{8}O$). i) How many electrons are present in each atom? ii) Write down their ground-state electronic configurations.

Ans10. i) Both atom contains 8 electrons but different numbers of neutrons, which makes $^{18}_{8}O$ heavier than $^{16}_{8}O$.

ii) Since the number of the electrons are the same in both atoms, their electronic configurations are identical: $^{18}_{8}O$: $1s^2 2s^2 2p^4$ and $^{16}_{8}O$: $1s^2 2s^2 2p^4$

Q11. Consider the electronic configurations of nitrogen with 7 electrons. Classify each of the given configurations into the following categories:

Ground: all electrons occupy the ground state.

Excited: one or several electrons occupy the excited states.

Impossible: the configuration is incorrect or cannot exist.

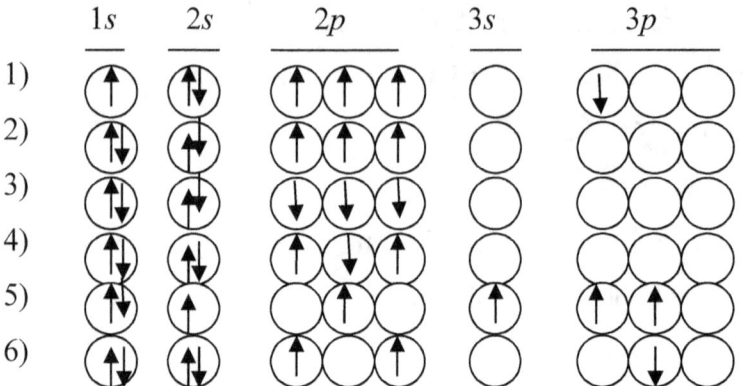

Ans11. The electrons in atoms or molecules occupy the ground states but some electrons can sometimes be excited from ground levels to excited levels (orbitals with higher energies). The diagonal arrow method is always useful to distribute the total number of electrons on the different orbitals. Next, the filling of each orbital (presented here as circles) with electrons must start with one electron in each orbital (or circle) with parallel spins (Hund's rule), then followed by a second electron with reverse spin until exhaustion of the electrons. Each orbital (circle) could have up to maximum of 2 electrons.

1) Impossible
2) Ground
3) Ground
4) Impossible
5) Excited
6) Impossible

Q12. i) Classify each of the given configurations into the following categories:

Ground: all electrons occupy the ground state.

Excited: one or several electrons occupy the excited states.

Impossible: the configuration is incorrect or cannot exist.

ii) In your view, why the last configurations are impossible? iii) Which neutral atoms could have each permissible configuration?

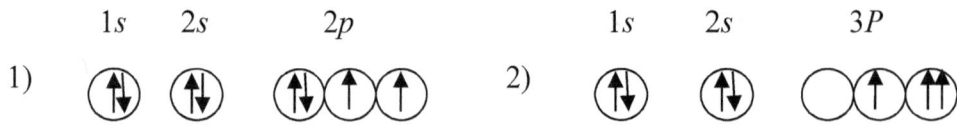

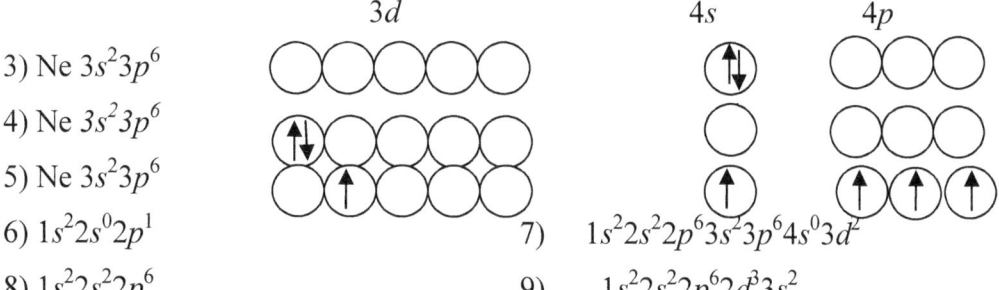

3) Ne $3s^2 3p^6$

4) Ne $3s^2 3p^6$

5) Ne $3s^2 3p^6$

6) $1s^2 2s^0 2p^1$

7) $1s^2 2s^2 2p^6 3s^2 3p^6 4s^0 3d^2$

8) $1s^2 2s^2 2p^6$

9) $1s^2 2s^2 2p^6 2d^3 3s^2$

Ans12. Look at the details of **Ans. 11**.

1) This configuration represents the ground state of an oxygen atom with 8 electrons. The Hund's rule of maximum multiplicity states that if several orbitals of same energy are available, the electron should occupy separate orbitals in unpaired states.

2) This configuration is impossible. It is the configuration of nitrogen atom with 7 electrons but because the distribution of the electrons in p orbital did not follow the Hund's rule, it becomes impossible.

3) The electrons in this configuration are in the ground state since both electrons occupy $4s$ orbital.

4) This configuration is impossible since it does not respect the Hund's rule.

5) Excited state since 1 electron in $4s$ orbital is excited to $3d$ orbital.

6) The configuration is in the excited state since the electron that should occupy the $2s$ orbital in the ground state is excited to $2p$ orbital.

7) Excited states since the two electrons from $4s$ are excited to the $3d$ orbital.

8) Ground state configuration because the orbitals are filled in increasing order of energy.

9) This configuration is impossible since electrons can not occupy $2d$ orbital with lower energy.

Q13. Determine the status of the following given species: neutral atom, cation or anion in the ground states, excited states, or impossible states.

1) $_3$Li $1s^2 2s^0 2p^1$

2) $_1$H $1s^2$

3) $_{16}$S $1s^2 2s^2 2p^6 3s^2 3p^6$

4) $_6$C $1s^2 2s^2 2p^2$

5) $_{10}$Ne $1s^2 2s^2 2p^7$

6) $_7$N $1s^2 2s^1 2p^3$

7) $_9$F $\quad 1s^22s^22p^53s^1$

8) $_2$He $\quad 1p^1$

9) $_{21}$Sc $\quad 1s^22s^22p^63s^23p^63d^14s^2$

10) $_8$O $\quad 1s^22s^22p^2$

Ans13. The periodic table gives the number of electrons of each species, which have to be compared to the given electrons provided for each case. If the numbers are equal, the species is neutral. However, if the number of electrons is higher or lower, the species becomes an anion or cation, respectively.

1) The species is neutral since Li has 3 electrons, which matches the given number. Li is in its excited state since the third electron that should occupy the 2s orbital is excited to 2p orbital.

2) H has 1 electron and here is given by 2 electrons, meaning that it gained one extra electron to form a negative ion (anion, H$^-$). The species is in the ground electronic configuration since the 2 electrons occupy the first orbital 1s.

3) S has 16 electrons but the given configuration has 18. Thus, s gained 2 electrons to become an anion S^{2-}. The configuration is in its ground state because lower orbitals in energy are occupied by the electrons.

4) Carbon has 6 electrons and the given configuration contained the same number of electrons. Thus, it is a neutral atom. All the electrons occupy the ground state.

5) Ne has 10 electrons but the given configuration contained 11 electrons. Thus, Ne received 1 extra electron to form Ne$^-$. However, the given configuration is impossible since the maximum electrons that could occupy the p orbital is 6 but here given by 7.

6) N has 7 electrons and the given configuration has only 6. Thus, it donated (or lost) 1 electron to form a cation N$^+$. The lost electron is removed from the orbital s to leave it with only 1 electron. We can consider the configuration as the ground state of the cation.

7) F has 9 electrons but the given configuration has a total of 10 electrons, meaning that it gained 1 extra electron to form F$^-$ (Ne configuration). Since the 2p orbital should be completed with 6 electrons before filling the 3s orbital, thus the configuration is in the excited state.

8) He has 2 electrons but in the given configuration has only 1. Thus it donated 1 electron to form a cation He$^+$. The given configuration is impossible since the lower orbital 1s should first be filled with electrons.

9) Sc has 21 electrons, which is also the case in the given configuration. Thus it is a neutral atom and the electronic configuration is in the ground state

10) O has 8 electrons but the given configuration has only 6. This means that it lost 2 electrons to form an anion O^{2-}. The configuration looks in the ground state since lower orbitals are first filled with electrons.

Q14. For each of the following species, determine the electronic configuration and predict their relative sizes: F^-, Na^+, Mg^{2+}, Ne, O^{2-}, and N^{3-}.

Ans14. F has 9 electrons and F^- gained 1 extra electron to reach a total of 10 electrons. Thus, the electronic configuration of F^- is $1s^2 2s^2 2p^6$.

Na has 11 electrons and lost 1 electron in Na^+. Hence, the configuration of Na^+ is: $1s^2 2s^2 2p^6$.

Mg has 12 electrons and lost 2 in Mg^{2+}. Therefore, the electronic configuration of Mg^{2+} is: $1s^2 2s^2 2p^6$.

Ne is a noble gas with 10 electrons (difficult to remove or add electrons). The electronic configuration of Ne is: $1s^2 2s^2 2s^6$.

O has 8 electrons but gained 2 extra electrons in O^{2-}. The electronic configuration of O^{2-} is: $1s^2 2s^2 2p^6$.

N has a total of 7 electrons but it gained 3 extra electrons in N^{3-} to form the electronic configuration: $1s^2 2s^2 2s^6$.

Because all these species (N^{3-}, O^{2-}, F^-, Ne, Na^+, Mg^{2+}) have the same total number of electrons (10 electrons), they are called iso-electronic species.

The size of iso-electronic species reduces as the atomic number or nuclear charge rises. Hence, the sizes of these species should decrease in the following order: $N^{3-} > O^{2-} > F^- > Ne > Na^+$.

Q15. Compare the electron affinity, electronegativity and ionization energy between each of the following atom pairs: Cu vs. Zn, K vs. Ca, S vs. Cl, H vs. Li, and As vs. Ge. Explain why.

Ans15. To be able to compare between these pairs, the electronic configuration of each of them should first be written.

$_{29}Cu$: $1s^2 2s^2 2p^6 3s^2 3p^6 4s^2 3d^9$

$_{30}Zn$: $1s^2 2s^2 2p^6 3s^2 3p^6 4s^2 3d^{10}$

$_{19}K$: $1s^2 2s^2 2p^6 3s^2 3p^6 4s^1$

$_{20}Ca$: $1s^2 2s^2 2p^6 3s^2 3p^6 4s^2$

$_{16}S$: $1s^2 2s^2 2p^6 3s^2 3p^4$

$_{17}$Cl: $1s^2 2s^2 2p^6 3s^2 3p^5$

$_1$H : $1s^1$

$_3$Li : $1s^2 2s^1$

$_{33}$As: $1s^2 2s^2 2p^6 3s^2 3p^6 4s^2 3d^{10} 4p^3$

$_{32}$Ge: $1s^2 2s^2 2p^6 3s^2 3p^6 4s^2 3d^{10} 4p^2$

The electron affinity relies on the energy released or consumed when 1 electron is added to a neutral atom. Zn has a *d* orbital fully filled with electrons, which makes it difficult to add extra electrons. By contrast, Cu only lacks 1 electron to complete its outer shell electronic structure. Thus, the electronic affinity of Cu is higher than that of Zn (Cu>Zn). The same scenario applies to (K>Ca). For S and Cl, since only 1 electron lacks to complete the *p* orbital of Cl versus 2 electrons for S, thus Cl should have more affinity to electrons than S (Cl>S). The addition of 1 electron to H yields the electronic structure of the stable noble gas He. Thus, H has more affinity to electrons than Li (H>Li). As has its *p* orbital half full, which gives it extra stability at the current state. The addition of 1 electron to Ge will induce a half full *p* orbital. Thus, Ge should have more affinity than As to one additional electron to make the *p* orbital half full and acquire more stability (As<Ge).

Electronegativity expresses the tendency of an element to attract electrons towards it in a chemical bond. Since Cu lacks only 1 electron to fill in its *p* orbital, it will strongly attract that electron to form a chemical bond with other species. Thus, the electronegativity of Cu is higher than that of Zn (Cu>Zn). By losing 1 electron, K will acquire the structure of the noble gas Ar with lower electronegativity in a chemical bond. Hence, the electronegativity of K is lower than that of Ca. By acquiring 1 electron, Cl will achieve the electronic structure of the stable noble gas Ar, thus its electronegativity is higher (Cl> S). The same scenario applies to H, which by acquiring 1 electron will have the configuration of the noble gas He. Therefore, its electronegativity is higher than that of Li which stabilizes by losing one electron to acquire the structure of He (H>Li). Finally, As with a half full *p* orbital is more stable and should have higher electronegativity in a chemical bond than Ge which could easily lose 1 electron (As>Ge).

The ionization energy refers to the energy required to liberate 1 electron. Because Zn has its *d* orbital saturated with electrons and pretty stable, thus removing electrons will be more difficult than in Cu with incomplete orbital (Cu<Zn). Ca has also a saturated orbital, hence more difficult to remove electron than in K (K<Ca). Cl lacks only 1 electron to acquire the Ar

structure, thus it has more tendency to attract electrons rather than giving them away. This makes its electronegativity higher (Cl>S). Li could easily lose 1 electron to acquire the He structure, making its electronegativity lower than that of H, which by gaining 1 electron acquires the He structure (H > Li). Finally, As is more structurally stable than Ge because its *p* orbital is half full. Hence, its ionization energy is higher (As > Ge).

Q16. Classify the following elements by increasing trend of first ionization energy, electron affinity, electronegativity, and atomic radii: C, O, and F. Explain why.

Ans16. To be able to explain the trends, the electronic configurations should first be written.

$_6C: 1s^2 2s^2 2p^2$

$_8O: 1s^2 2s^2 2p^4$

$_9F: 1s^2 2s^2 2p^5$

The difference between these elements lies in the number of electrons occupying the last *p* orbital, which would mostly define their chemical properties, such first ionization energy, electron affinity, electronegativity, and atomic radii.

The first ionization energy expresses the energy required to liberate 1 electron. F requires only 1 extra electron to acquire the structure of the noble gas Ne. Thus, its ionization energy should be higher than that of oxygen and carbon requiring 2 and 4 extra electrons, respectively. Hence, the ionization energy should increase in the following order: C < O < F.

The electron affinity refers to the energy released or consumed when putting 1 extra electron on a neutral atom. F has more affinity to gain 1 electron to acquire the structure of the noble gas Ne, followed by O and C (C < O < F).

The electronegativity defines the tendency of an atom to attract electrons to its side when put in a chemical bond. F is more electronegative than O and C because it requires only 1 extra electron to acquire the structure of the noble gas Ne. Thus, electronegativity increases from C to F: C < O < F.

The atomic radii is influenced by the number of protons (nucleus) and electrons. The more the number of protons is high, the more the electrostatic force exerted by the protons on the electrons is high. This contracts the atomic radii. Since F has higher number of protons than O and C, it should have the smallest radii: C > O > F

Q17. What is the number of unpaired electrons in the ground states of the following species: C^{-4}, F^-, and Ne.

Ans17. Carbon has 6 electrons and gained 4 in C^{-4}. F has 9 electrons and gained 1 in F^-. Ne has a total of 10 electrons. Thus, these species are isoelectric since they have a similar number of electrons. Their electronic configurations are:

	1s	2s	3p	
C^{4-}	(↑↓)	(↑↓)	(↑↓)(↑↓)(↑↓)	Zero unpaired electrons
F^-	(↑↓)	(↑↓)	(↑↓)(↑↓)(↑↓)	Zero unpaired electrons
Ne	(↑↓)	(↑↓)	(↑↓)(↑↓)(↑↓)	Zero unpaired electrons

All the species have zero unpaired electrons.

Q18. According to the periodic table, what is the number of protons, neutrons, and electrons in nitrogen 14? How is this atom influenced by removal of: 1 neutron, 1 proton, 1 electron, and (1 proton + 1 electron)?

Ans18. N (14) has 7 protons, 7 neutrons, and 7 electrons.

Its symbol is: $^{14}_{7}N$, where 14 correspond to the number of (protons + neutrons) and 7 is the number of electrons.

Removal of one neutron will have no effect on the proton or electrons.

Removal of one proton will form a negatively charged anion since the number of electrons is higher than the protons. The atomic radii will eventually increase since less electrostatic force is exerted from the nucleus on the electrons.

Removal of 1 electron will form a positively charged cation since the number of protons is higher than that of the electrons. The atomic radii will eventually retract since there are too many protons for fewer electrons.

The removal of (1 electrons, 1 proton) will form carbon but with 1 extra neutron in its nucleus.

Q19. The molecular formula of oxalic acid is $C_2O_4H_2$. Calculate its molecular mass and the number of grams of carbon present in 100g, 80g, and 920g of oxalic acid.

Ans19. The molecular mass of a compound is obtained by adding the molar masses of all atoms forming the compound. Hence, M ($C_2O_4H_2$) = 2×M(C) + 4×M(O) + 2×M(H) = 2×(12) + 4×(16) + 2×(1) = 90 g mol^{-1}

90 g of oxalic acid contains 24 g carbon. Hence, 100 g oxalic acid will contain: $\left(\frac{24}{90.03}\right) \times 100 = 26.66$ g of carbon.

80 g of oxalic acid will contain: $\left(\frac{80}{90.03}\right) \times 100 = 21.33$ g carbon.

920 g of oxalic acid will contain: $\left(\frac{920}{90.03}\right) \times 100 = 245.27$ g carbon.

Q20. i) Which of the following elements has more atoms: a) 1 g Al, 1g Au or 1 g H? ii) Estimate the number of atoms of each element present in 1 g.

Ans20. i) The molar mass of each element can be found in the periodic table. For Al, 1 mole weights 27 g, and the number of atoms in 27 g equals the Avogadro's number (6.02×10^{23}). Hence, 1 g Al should have: $\frac{6.02 \times 10^{23}}{27} = 2.23 \times 10^{22}$ atoms.

For Au, 1 mole weights 197 g and contains 6.02×10^{23}. Thus, 1 g Au will have: $\frac{6.02 \times 10^{23}}{197} = 3.05 \times 10^{21}$ atoms.

For H, 1 mole weights 1 g and contains 6.02×10^{23} atoms. Therefore, 1 g H will have: $\frac{6.02 \times 10^{23}}{1} = 6.02 \times 10^{23}$ atoms.

The number of atoms in 1 g of each element increases in the following order: Au<Al<H.

Q21. Calculate the number of atoms present in 0.001857 g of tungsten.

Ans21. The molar weight of tungsten (W) is 183.85 g mol^{-1}, which contains 6.02×10^{23} atoms. Hence, 0.001857 g of W contains: $\frac{0.001857 \times 60.2 \times 10^{23}}{183.85} = 6.07 \times 10^{18}$ atoms

Q22. i) Write down the chemical formulas of the following compounds: lithium chloride, zinc sulfate, calcium sulfate, copper hexacyanoferrate(III), ferric chloride, and chromic fluoride. ii) What features share these compounds?

Ans22. i) The respective formulas are: LiCl, $ZnSO_4$, $CaSO_4$, $Cu_3[Fe(CN)_6]_2$, $Fe_2(Cl)_6$, and CrF_3.

ii) All these compounds are in the salt or complex forms.

Q23. i) Calculate the molecular weight of $KMnO_4$. i) What is the weight percent of each element in $KMnO_4$?

Ans23. i) The molecular weight of $KMnO_4$ is calculated by adding the weights of all the atoms forming $KMnO_4$. The molecular weight of $KMnO_4$ = M(K) + M(Mn) + 4 × M(O) = 39.1 + 54.9 + 64 = 158 g mol^{-1}

ii) The weight percent of K in $KMnO_4 = \left(\frac{39}{158}\right) \times 100 = 24.75\%$

The weight percent of Mn in $KMnO_4 = \left(\frac{54.9}{158}\right) \times 100 = 34.75\%$

The weight percent of O in $KMnO_4 = \left(\frac{64}{158}\right) \times 100 = 40.5\%$

The addition of all weight percent gives 100%.

Q24. What would be the empirical formula of cement containing 52.7% calcium (Ca), 12.3% silicon (Si), and 35.0% oxygen (O)?

Ans24. The mass of the constituents should first be converted into moles of constituents by dividing by their atomic masses. Ca = $\frac{52.7}{40.08}$ = 1.31 moles, Si = $\frac{12.3}{28.09}$ = 0.44 moles, and O = $\frac{35}{16}$ = 2.18 g moles.

The atomic ratio is obtained by dividing the moles of each constituent by the smaller number of moles (0.44 moles). Hence, Ca:Si:O = $\frac{1.31}{0.44}:\frac{0.44}{0.44}:\frac{2.18}{0.44}$ = 3 : 1 : 5

The empirical formula for the proposed cement is Ca_3SiO_5.

Q25. A metal oxide X_2O_3 contains 68.4% metal by weight. What is the atomic weight of X and its chemical formula?

Ans25. The atomic weight of X = $\left(\frac{2X}{2X+3(16)}\right) \times 100$ = 68.4. This gives a value of M (X) = 51 g mol^{-1}. According to the periodic table, X = V (vanadium), and the chemical formula is V_2O_3.

Q 26. A compound AB is hypothetically formed by reaction between 10 moles of A with 30 moles of B. Another compound AC_2 is formed by reaction of 6 moles of A with 36 moles of C. i) Assuming that the molar weight of element B is 60 g mol^{-1}, estimate the molar weights of both elements, A and C. ii) Could these reactions practically occur and why?

Ans26. i) The stoichiometry of the first reaction indicates that 1 mole of A reacts with 3 moles of B and since the molar weight of B is 60 g mol^{-1}, thus the molar weight of A = $\frac{60 \times 10}{30}$ = 20 g mol^{-1}.

The stoichiometry of the second reaction suggests that 1 mole of A reacts with 3 moles of C and since the molar weight of A is 20 g mol^{-1}, hence the molar weight of C = $\frac{36 \times 20}{6 \times 2}$ = 60 g mol^{-1}.

The periodic table indicates that the element with the molar weight of 20 g mol^{-1} corresponds to the noble gas Ne and 60 g mol^{-1} does not exist but the closest elements with this molar weight are Co or Ni with 59 g mol^{-1}. ii) Since Ne is a very stable element (noble gas), it is unlikely to react with any other element. Overall, these reactions are just hypothetical and will not happen practically.

Q27. Are the following given reactions balanced? If not balance each of them and explain how.

1) $Na_2SO_3 + HCl \rightarrow NaCl + SO_2 + H_2O$

2) $Mg_3N_2 + H_2O \rightarrow Mg(OH)_2 + NH_3$

3) $Pb + PbO_2 + H_2SO_4 \rightarrow PbSO_4 + H_2O$

Ans27. A chemical reaction is balanced if the mass and charge on each side of the equation are the same. All the given reactions have zero charges on both sides, meaning that they are charge balanced. In terms of mass, each element should have the same number of atoms on each side of the equation, which is not the case for each of the given reactions. The simplest way to balance these reactions is to multiply by factors to equalize the number of atoms present on each side of the reaction.

The balanced equations are:

1) $Na_2SO_3 + 2HCl \rightarrow 2NaCl + SO_2 + H_2O$

2) $Mg_3N_2 + 6H_2O \rightarrow 3Mg(OH)_2 + 2NH_3$

3) $Pb + PbO_2 + 2H_2SO_4 \rightarrow 2PbSO_4 + 2H_2O$

Q28. The reaction between ammonia and oxygen forms nitric oxide (NO) and water, and the reaction between nitric acid with zinc hydroxide ($Zn(OH)_2$) produces zinc nitrate ($Zn(NO_3)_2$) and water. Write down each reaction, making sure that they are balanced.

Ans28. Ammonia has the formula NH_3, oxygen gas exists in the form of O_2. Thus, the first reaction gives: $4NH_3 + 5O_2 \rightarrow 4NO + 6H_2O$

Similarly, nitric acid has the formula HNO_3. Hence, the second reaction gives:

$Zn(OH)_2 + 2HNO_3 \rightarrow Zn(NO_3)_2 + 2H_2O$

These reactions are mass and charge balanced.

Q29. The burning of carbon in air forms carbon dioxide. Estimate the weight of the resulting carbon dioxide when 50 g of carbon is burned in air.

Ans29. Air contains mainly oxygen and nitrogen. Burning processes require oxygen. Thus, by assuming carbon in its elemental form (C), the burning reaction can be written as:

$C + O_2 \rightarrow CO_2$

The stoichiometry of the reaction indicates that 1 mole of C requires 1 mole of O_2 to produce 1 mole of CO_2. The number of moles corresponding to 50g C is: $\frac{50}{12}$ = 4.16 moles, where 12 represent the molar mass of C. Since 1 mole C gives 1 mole CO_2, thus 4.16 moles of CO_2 are produced. This corresponds to 4.16 × 44 = 181.28 g, where 44 is the molar mass of CO_2. In sum, the reaction produces 181.28g of CO_2.

Q30. i) Is the reaction between potassium dichromate and both oxalic acid and sulfuric acid balanced?

$3 H_2C_2O_4 + K_2Cr_2O_7 + H_2SO_4 \rightarrow 2KHSO_4 + Cr_2(SO_4)_3 + 6CO_2 + 7H_2O$

ii) A potassium dichromate solution (460 ml, 0.100 M) reacts with excess oxalic and sulfuric acids. Estimate the number of moles and grams of the formed CO_2.

Ans30. i) No, the reaction is not balanced. The charge is balanced but the mass is not. This could be balanced by multiplying the reaction by factors to equalize the number of atoms on each side of the equation. We start with the less abundant elements, sulfur in this case. By multiplying H_2SO_4 by a factor of 5, the reaction becomes balanced.

$3H_2C_2O_4 + K_2Cr_2O_7 + 5H_2SO_4 \rightarrow 2KHSO_4 + Cr_2(SO_4)_3 + 6CO_2 + 7H_2O$

ii) The stoichiometry of the reaction indicates that 1 mole $K_2Cr_2O_7$ forms 6 moles CO_2. Since 0.100 M $K_2Cr_2O_7$ is used in presence of excess of other compounds, all $K_2Cr_2O_7$ should react to form 0.6 M CO_2. This is the equivalent of $0.6 \times 44 \times 0.46 = 12.14$ g, where 44 is the molar weight of CO_2 and 0.46 L is the volume of the solution.

Q31. The reaction between vanadium oxide (VO) and iron oxide (Fe_2O_3) forms V_2O_5 and FeO. i) Write down the balanced reaction. ii) Estimate the number of moles and grams of V_2O_5 when 6.50 g VO react with excess Fe_2O_3. How many grams of V_2O_5 are produced? iii) Calculate the number of moles and grams of formed V_2O_5 when 4.00 g VO react with 11.5 g Fe_2O_3.

Ans31. i) Assuming no other reactants or products are involved, the balanced reaction can be written as: $2VO + 3Fe_2O_3 \rightarrow V_2O_5 + 6FeO$

ii) The stoichiometry of the reaction indicates that 2 moles VO form 1 mole V_2O_5. Since excess $3Fe_2O_3$ is used, all 6.5 g should react to completion. The number of moles of VO corresponding to 6.5g is: $\frac{6.5}{66.94} = 0.097$ moles, where 66.94 represents the molar weight of VO. Thus the number of moles of the resulting V_2O_5 is: $\frac{0.097}{2} = 0.048$ moles of V_2O_5. This is equivalent to: $0.048 \times 182 = 8.83$ g, where 182 g mol^{-1} is the molar weight of V_2O_5.

iii) The reaction of 4 g VO with 11.5 g Fe_2O_3 means that the reaction will be limited by the 11.5 g Fe_2O_3 because its stoichiometry is higher. 11.5 g Fe_2O_3 corresponds to: $\frac{11.5}{159.688} = 0.072$ moles of Fe_2O_3, where 159.688 is the molar mass of Fe_2O_3. Since 3 moles of Fe_2O_3 produce 1 mole of V_2O_5, thus 0.072 moles of Fe_2O_3 produce: $\frac{0.072}{3} = 0.024$ moles of V_2O_5. This is equivalent to: $0.024 \times 181.88 = 0.024$ g of V_2O_5.

Q32. i) If an electron in a hydrogen atom is characterized by the principal quantum number $n = 3$. Estimate the values of the second quantum number l.

ii) Calculate the values of the magnetic quantum number m for an electron with $l = 3$.

iii) Estimate the values of the quantum numbers n, l, m and s for an electron in the $3d$ orbital.

Ans32. i) l is the azimuthal quantum number characterizing the shape of the orbital and its angular momentum. l could have integers from 0 to $(n - 1)$. Thus, for $n = 3$, the values of $l = 0, 1$, and 2.

ii) m is the magnetic quantum number with integer values ranging from $-l$ to l, representing specific orbital within a subshell. For $l = 3$, the values of $m = 0, \pm1, \pm2, \pm3$.

iii) An electron in the $3d$ orbital would have the following quantum numbers: $n = 3$, $l = 2$, $m = (0, \pm1, \pm2)$, and $s = \pm1/2$.

Q33. i) Estimate the radius of the fourth Bohr orbit of hydrogen atom if the first Bohr orbit has a radius of 0.529 Å. ii) What would be the radius of boron (B) under the same conditions?

Ans33. i) Since the hydrogen atom has $Z = 1$, the radius can be calculated by the relationship: $r = 0.529 \times \frac{n^2}{Z} = 8.4$ Å.

ii) The difference between H and B lies in the Z number. For Z (B) = 5, $r = 0.529 \times \frac{n^2}{Z} = \frac{8.4}{5} = 1.68$ Å.

Q34. Consider two hydrogen atoms (A and B) with 1 electron each. The electron occupies the first Bohr orbit ($n=1$) in atom A and the fourth Bohr orbit ($n = 4$) in atom B. Which of the two atoms A or B: i) should have the excited state electronic configuration, ii) faster speed of the electron, iii) larger radius, iv) lower potential energy, v) and elevated ionization energy?

Ans34. i) Atom A is in the ground state and atom B in the excited state. ii) The radius of the moving electron is inversely proportional to the square of its velocity. Therefore, the electron in the ground state should move faster. iii) The atom B with an electron in the excited state has a larger radius ($n = 4$). iv) The excited electron in atom B should have lower potential energy since it is already in the excited state. v) The electron in atom A requires higher ionization energy since it occupies a stable state.

Q35. Determine the number of unpaired electrons in: Cr^{3+}, Cr^{2+}, Cr^{5+}, and Cr^{6+}. What would happen to these species in the excited states?

Ans35. To determine the number of impaired electrons, the electronic configurations of the species should first be written.

Cr has a total of 24 electrons and Cr^{3+} lost 3 electrons, thus 21 electrons: $1s^22s^22p^63s^23p^64s^23d^1$.

Cr^{3+} has one unpaid electron in the ground state. At the excited state, 1 electron of the 4s will be excited to the d orbital. The three unpaired electrons could combine with electrons of other species to form chemical bonds, such as during the formation of complexes.

Cr^{2+} will have 2 unpaired electrons in the ground state and 4 in the excited state.

Cr^{5+} will have 1 unpaired electron in the ground state, and 1 or several at the excited state, depending on which electrons are excited.

Cr^{6+} will have 0 unpaired electrons in the ground state, and several in the excited state, depending on which electrons are excited.

Q36. Using the electronic configuration explain why the metals (Cu, Ag, Au) have high electric conductivities.

Ans36. The electronic configurations of Cu, Ag and Au are as follows:

Cu $\quad\quad\quad 1s^22s^22p^63s^23p^64s^13d^{10}$

Ag $\quad\quad\quad 1s^22s^22p^63s^23p^64s^23d^{10}4p^65s^14d^{10}$

Au $\quad\quad\quad 1s^22s^22p^63s^23p^64s^23d^{10}4p^65s^24d^{10}5p^66s^14f^{14}5d^9$

All these metals share similar features in terms of the presence of one single valence electron in the s orbital ($4s^1$, $5s^1$, $6s^1$). This electron can freely move to cause strong repelling reaction on the other electrons. A single valence electron moving with little resistance leads to elevated electric conductivity.

Table of Content

Discount offers	1
Introduction	2
Abstract	3
1. Atoms as building blocks of matter	3
1.1. Matter and substances	3
1.2. Elements and structure of atoms	4
1.3. Some characteristics of atoms	4
2. Periodic table of elements	5
3. Basic electronic models of atoms	7
4. Electronic structure	8
4.1. Quantum numbers	8
4.2. Energy levels and determination of electronic configuration	9
Summary	11
References	11
Practical Questions/Problems with Solutions	13
Table of content	30
About the author	32

www.ingramcontent.com/pod-product-compliance
Lightning Source LLC
Chambersburg PA
CBHW062236220526
45471CB00009B/3512